Blacksmithing Book for Beginners

Learn How to Forge 15 Easy Blacksmith Projects
with Step By Step User Guide Plus Tips, Tools and
Techniques to Get You Started

By

Luke Wade

Disclaimer

This publication is designed to provide competent and reliable information regarding the subject matter covered. However, the views expressed in this publication are those of the author alone, and should not be taken as expert instruction or professional advice. The reader is responsible for his or her own actions.

The author hereby disclaims any responsibility or liability whatsoever that is incurred from the use or application of the contents of this publication by the purchaser or reader. The purchaser or reader is hereby responsible for his or her own actions.

Table of Contents

Introduction

Metal production is a fast-growing business and industry. It has gone through many stages of advancement, from rough and rugged to easy and classic. One of those rough and rugged forms of metal production that has been around for donkey years is Blacksmithing.

Blacksmithing is an old craft, one of the oldest craft man has ever invented and preserved for years. It is almost as old as agriculture!

Starting in an ancient city by local settlers, it began to spread out into other countries and became a worldwide craft for making metallic tools.

Oh well, I can't really say what was used in place of all the hand tools we have now, but blacksmithing exposed the possibility of making tools from metals and gave importance to unrefined steels.

Well, this has made our lives easier in many ways. Thanks to the early blacksmiths who gave themselves to this discovery.

Blacksmithing, as a craft, can be very technical and stressful. The process requires a degree of smartness, patience, technicalities and creativity. Above all of these, blacksmithing is a highly creative art.

The ability to create has always been in man. Yes, everyone possesses the ability to create and recreate, including you.

Time has proven this! Right from the medieval time when man started producing house tools and utensils from unrefined steels until date, man's creative ability has never been more glaring! Just in case you doubt that. Creativity lives in man.

Not only is it a creative art, but a very inspirational one at that. The stages and processes involved in blacksmithing are also quite inspirational that wise life quotes have been generated from it. Quotes like "not all that glitters is gold" and "strike when it is hot." Smiths worldwide were viewed in medieval times as very honorable men and acknowledged as such to date. In some places in medieval times, they were adored and worshipped. Indeed, it was a deeply honorable art.

Old blacksmiths had several techniques for making metallic products, some of which are still in use today.

Some other complex techniques have also been introduced, and we will be discussing each of these techniques in detail, as they determine to a large extent the outcome of your blacksmithing exercise.

The blacksmith has his basic tools of operation like anvil, forge, hammer, chisels, punches, and quench tub. These are very important tools that he cannot do without. This book teaches the use of all these tools and their importance in the blacksmithing process.

The blacksmith's workshop is his sacred shrine where he carries out his operations and should be treated as such.

As a beginner, there are so many smithing codes you will need to get familiar with before you get started. Passion is not enough to make you a blacksmith, neither does a workshop or tool bag. There are basic systems of operation that you need to know by heart before you can make a career of blacksmithing.

This is why I celebrate you for buying this book. Your journey has just started.

This is why we need to preserve this heritage and continue to pass it on. Time and technology must not and cannot steal what our ancient fathers spent their lives building and handed to us; instead, we have to use them as an advantage to build our own career in blacksmithing.

Although the old and crude methods of blacksmithing were highlighted in this book, you can take advantage of the resources that technology has availed us in our time.

There are so many innovations that have been made in place of the crude ones. There are more technical tools that help in forging, shaping and cutting. Hence, making it easier for you to achieve a good result as you would have with a machine. It also makes your work smoother and helps you keep a tidier workplace. The old and conventional way of blacksmithing can be very stressful and demanding.

Today, it is easier and faster for you to make a singular metal than ever. Using these modern tools, you can make in a few hours what would have normally taken all of your time. These tools accelerate the smithing process. However, modernization of tools or not, you can never replace hitting the steel.

Trading as a blacksmith in this time can be quite challenging and difficult. It could seem almost impossible to get sales, especially with machines fully taking over the art of metal production. Nonetheless, it doesn't nullify the relevance of the art at all. Instead, it allows you to prove the uniqueness of the craft.

Keep your passion intact and get ready to journey back and forth into the world of blacksmithing as I expose to you the basics of this incredible art.

Chapter 1

Essentials of Blacksmithing

Do you think that blacksmithing has become obsolete and blacksmiths no longer exist? How wrong you are. The art of blacksmithing has been around for centuries and will continue to forge on as long as people hold on to it and pass down the long-held traditions of this craft.

However, technology seems to have overridden this art and swept it under the carpet. The highly reverenced art of blacksmithing that happened to be one of the most popular trade skills around the 13th century down to the 18th century is now regarded as obsolete due to technology. Technology has introduced sophisticated machines in replacement for manual labor and these machines have proven to be more efficient, hence reducing the demand for blacksmiths.

The decline in the relevance of blacksmithing is owing to a factor bigger than time. Aside from the mechanization of the art, most young persons in the recent generation have gone on to learn digital and other technical skills that could earn them a place in a cozy office or bring good contracts without much hassle. It is very rare to see a young person who is interested in crude skills like blacksmithing today.

However, the emergence of technology has not totally burnt out crude skills such as blacksmithing. Blacksmithing is a unique ancient art that cannot be buried as long as there are people who are passionate about it. Moreover, certain arts can only be created manually by a blacksmith.

To this generation and beyond, blacksmithing will forever remain relevant. However, we need to explore and educate ourselves right from the foundation level of what blacksmithing is all about to preserve this heritage.

In this chapter, we will be taking a slow and steady ride to understand the meaning of blacksmithing, the history of blacksmithing, how it works, and ways of modernizing this trade.

What is Blacksmithing?

Blacksmithing can be defined as a crude art, skill, craft, and trade of fabricating objects from iron or steel by forging: hammering, bending, and cutting the metal.

At a basic level, forging means to form a metal object (iron or steel) by heating it in a direct fire or furnace and hammering it into shape. Blacksmiths heat it into a malleable form

This is simply what blacksmithing is about; the making of metal through forging. From our definition of blacksmithing, you will agree that it is a craft that involves high energy. This explains why the craft is mainly dominated by men. In comparison with the operation of a whitesmith, it is called 'a heavy work.'

Whitesmith is also a metalworker who works on white or light-colored metals. They mostly do the finishing work of polishing, filing, and lathing on iron and steel. Other craftsmen work with metals; some of them are called farrier, silversmith, shoer, and wheelwrights. Just like the whitesmiths, they work on special metals. They are specialized metalworkers.

Blacksmithing involves the crude work of forging raw metal (iron and steel) through heat into a malleable form and beating them into shape using hand tools. It involves the special ability to conjure objects, tools, and weapons from crude iron ore. The person who does blacksmithing is popularly called a blacksmith or an ironsmith.

Below is a list of things a blacksmith makes:

- Hardware: hand nails for building and construction, door and drawer, metal hooks, door hinges, screws, bolts, locks, latches, railings, gates, and nuts.
- Tools: hammers, cutlasses, shovels, axes, hoes, and other agricultural implements.
- Armor: helmets, shields, armor.
- Weapons: knife, javelin, swords, spear, clubs, axes, and other crude weapons.

Blacksmithing also involves the knowledge and ability to repair metallic objects like agricultural implements. A place where a blacksmith works is called a smithy, a forge, or a blacksmith shop.

Most of the blacksmith's tools are heavy and dangerous to be moved about; this is why they must be kept carefully in a workshop.

Indeed, blacksmithing is an age-long skill of making and repairing metallic objects through forging, cutting, and shaping. The blacksmiths were regarded as the king of trades because of their ability to create their own tools and for others to use. Blacksmithing is the beginning and foundation of metal production. It still remains the elementary of metal production.

History of Blacksmithing

The origin of blacksmithing can be traced to different dates, in different geographical locations.

However, if we are to trace the history of blacksmithing worldwide, we are to start from the Mesopotamian people's story. History has it that the first metalworking was discovered as early as 4,500 BC in Mesopotamia. The villagers, also known as settlements, began using firewood and rocks to heat up metal pieces like copper into desired shapes for hand tools and weapons.

In a short time, other metals like gold, silver, tin, lead, and iron were discovered and the settlements began using these stronger metals (iron and steel) to make hand tools and stronger weapons for war.

By 1200 – 1000 BC forging had become part of human civilization and that was when the term blacksmith emerged. From 1000 BC, the forging techniques began to spread throughout Europe and the rest of the old world.

However, in Egypt, which happened to be a world power, the first art of blacksmithing was discovered in 1350 BC. A group of men that turned iron into a dagger was found in Hittite. Then, the Hittites invented the art of forging and tempering. They worked secretly and hid their ironworking skills from every other person.

However, in 800 – 500 BC when they became scattered to Greece and Balk, their ironworking secrets were unveiled and it began to spread through Asia, especially Greece.

Most smiths around this time forged iron using charcoal fire from wood (they discovered that wood fires were hotter and had a higher intensity because of the air blast and the charcoal) and bellows to increase the heat of the fire and a hammer or stone to bend, cut and shape the metal into different tools, and weapons.

They were forging metals directly in the fire. It was crude at the time and only involved fire and basic tools. However, it sufficed to create simple weapons like spearheads and arrows.

In the middle age (5th to 15th century), during the onset of the industrial revolution, the smiths began to specialize under different branches.

There was the whitesmith who worked on white metals like lead, and the blacksmith who worked on raw iron, the farrier who made horseshoes, the goldsmiths, locksmith, nailsmiths, and others. Smithing became an industry on its own.

People began attaching smiths, miller, and cooper to their names or their children's names, to show their connection to the profession. It became a heritage parents handed down to their children. For many families, it was a family trade. All the male children were introduced to the craft while the women worked the garden and prepared meals for the whole family. Before blacksmithing became a popular business craft, most smiths made tools for personal and family use.

Then, many smiths used their backyard and empty spaces in their house as a workshop.

A picture of a man and his family in his workshop

By the 16th century, it was obvious that blacksmithing is a household craft that has come to stay, as there was a higher demand for metallic objects and tools. The government began to hire smiths to make weaponry and armory for the army. Also, some household companies demanded for stronger steels and metals. The need for these things became more urgent worldwide even as civilization started to spread. Everyone needed a gate for their house, hinge for their door, nut for their hook and knives amongst many other things.

The metal was a basic household and industrial need, hence the advancement in the craft. Around that time, there was a high boost in the business as many more demands were pouring in by the day.

It took centuries before blacksmiths were able to create stronger and harder metals; there were a lot of experiments before the invention fully manifested. Blacksmiths spent years researching ways to modify the carbon content of iron to be used for multiple purposes. Eventually, it was discovered that by smelting iron with carbon, a stronger metal was produced. The practice began and spread throughout the whole of Europe and some western countries.

One result of advancement in blacksmithing was the emergence of Decorative blacksmithing, which was introduced by Jean Tijou in the 17th century. This art required a level of creativity beyond the ability to beat metal in heat and shape it into something tangible. A lot of blacksmiths launched into this form of blacksmithing and began to produce decorated blacksmithing art products.

Some of them were images of people, animals, and natural things like flowers. They were designed in a very artistic form. Unlike the usual blacksmithing, which was to make a product for use, decorative blacksmithing was to produce a product for interior or exterior decoration.

Most blacksmiths began customizing and designing their products to look attractive and not just plain. Instead of a normal plain straight gate, we now have flowered gates with different designs.

Example of decorative blacksmithing

At this point, metal production and blacksmithing had spread throughout the Asian countries and had become a part of everyday life. Blacksmiths became respected artisans.

This led to the iron age in 800 – 500 BC. The iron age represented a boom in the market of smiths and a lot of entrance into the field.

During this time, bloomer furnaces started to replace charcoal fires. This furnace was made of clay and stone that could take repeated heat; it was designed to be heat-resistant. Pipes referred to as tuyeres, were built into the furnace and used to force air into the furnace using bellows to heat up the charcoal and increase the furnace temperatures.

The bloomer eventually wiped out charcoal fires after the discovery of water power. Water was discovered as an effective way to power bloomeries around the 13th century.

Waterwheels were used to power the bellows and allow for bigger and hotter bloomer furnaces with an improvement in forging production. The bellow or a blower is used to concentrate air and heat the forge to temperatures of 2000-3000*. The steel ought to be heated to around 2000 degrees Fahrenheit

After the 19th century, there was a strong turning point in the blacksmithing industry as there was a discovery of something known as modern-day forging techniques.

Francois Bourdon and James Nasmyth developed a tool known as the steam hammer. It is an industrial power hammer that is driven by steam and used for shaping and forging. This concept was first described by James Watt around the year 1784. However, it was later in 1840 that the machine was invented. James Nasmyth and Francois Bourdon had an acrimonious dispute in 1843 over who invented the machine. Eventually, the patent was handed over to James Nasmyth.

The steam hammer is still used today along with electric powered units.

A picture of a steam hammer

This made forging a lot easier. Thousands of years after, the process of forging still remains the most effective means of metal production and is practiced worldwide. With the emergence of technology, there is a higher demand for metal production and metallic products. Today, with computers, one could automate the forging process and streamline the process, making quick and precise parts for industries like aerospace and transportation.

Unlike before, when blacksmithing was a family craft and heritage, today, there are associations and organizations of the blacksmith in almost every location.

Today, blacksmithing is taught in boot camps, teenager summer camos, and secondary schools of developed countries. Volunteer smiths mostly organize these training pieces in local towns to ensure that the heritage of blacksmithing is passed down to the coming generation and the craft doesn't go into extinction. It is amazing that individuals have taken it upon themselves to pass down the knowledge of this great skill by creating free opportunities for young people to learn one of the longest established craftsmanship discovered. This craft will continue to thrive and exist over the years.

How Does Blacksmithing Work?

We have described some processes of blacksmithing in discussing the history of blacksmithing.

Despite the advancement in technology, many of the earliest techniques are unchanged, even in modern blacksmithing. There are four basic processes of blacksmithing: heating, holding, hitting and shaping.

Heating

It begins with heating pieces of wrought iron or steel until it becomes soft enough to be shaped with tools like hammer, chisel or an anvil.

You would also need to use a quenching bucket to cool the metal. Mineral oils are often used to facilitate the hardening of steel by controlling heat transfer and prevent distortion or cracking.

Finally, you'll need a safety apron and safety gloves. Although most persons use their bare hands, you wouldn't want to smear hot metal on yourself as a beginner mistakenly. Therefore, a protective glove and apron are very necessary.

Image showing the process of heating

Hitting

Hitting in blacksmith must not just be forceful but accurate. The metal piece must be hit at the right place with a sufficient level of force. The blacksmith needs certain tools that can help achieve a perfect level of accuracy.

- Anvil: This is the tool placed under the metal the blacksmith is hitting. It helps absorb the pressure from the blacksmith's hammer to prevent the metal from disengaging and breaking. It also helps to balance the force from the hit and rebound it through the metal, making the work less strenuous. It is majorly for stabilization and can hold heavy tools. This tool is made of cast steel.

- Hammers: The hammers come in various weight, head styles, and shapes. The type of hammer to be used is determined by the type of metal the blacksmith is beating and the style of decoration he intends to make. The hammer is used to hit the metal to control the metal volume and flatten it.

Image of an anvil and two different hammers

Holding

There are different tools used to hold the metal during the heating and shaping process. Of course, each of these materials is for special purposes.

Firstly we have tongs. Tongs are used to pick hot pieces of metal from the forge. It is made of wrought iron and has a large jaw that can lift any size of metal. Its jaw is also flat to prevent it from scratching the work.

Next are vices and clamps: these are powerful tools that can withstand intense pressure. It is used to firmly hold hot iron while it is hammered, chiseled, or twisted. It has been proven to last decades despite the kind of pressure it passes through.

Image of a Tong

Shaping

Blacksmithing is not complete without the shaping process. It is necessary to adjust the metal into the desired shape. You can't beat metal into shape; there is a need to shapen it using appropriate tools. This is still performed using a hammer but is done more carefully.

A picture showing the shaping process of a metal

At this point, you will agree with me that blacksmithing is not as complex as it seems. All you need is the right knowledge and adequate tools.

Modernizing Your Blacksmithing Business

Just as we have said earlier, technology has increased the demand for blacksmithing and metal production. It has also provided advanced and sophisticated equipment for blacksmithing.

However, some persons still choose to use the crude and old-fashioned way of doing it.

Suppose you would actually want to modernize your blacksmithing business. In that case, you should consider adopting the new blacksmithing technique while holding on to the basic practices and processes we have listed above: heating, holding, hitting, and shaping.

In as much as blacksmithing appears to be replaced by sophisticated machines and computers, there is still a market for anyone passionate about blacksmithing. This is why you have to modernize your business to stay on top of your game. The benefits of modernizing your blacksmithing business are immeasurable. It helps you save energy and time. By this, it enables you to meet up to deliveries in time and produce a more perfect work.

With modern tools, you can create a perfect hand tool or hardware without much hassle and in a shorter time. Also, there are designs the local method of blacksmithing cannot create because it is archaic, but by adapting to the new method, you are expanding your shores.

Technology has set the new standard for blacksmithing. If your smithing business will thrive in this time, you have to adopt the standards.

Below are listed tips to help you start your journey to modernization

- Get modern forging tools
  Today with machines, advanced metal production industries can create multiple products in a short time. However, this multiplicity advantage cannot be placed over the uniqueness of a metal produced by a professional blacksmith. Imagine how much you can achieve when you enhance your unique skill with modern tools. There's nothing like the master

doing his craft with his hands. The ability to create tangible materials from iron ore is almost magical and something that technology can never take away. However, you can't control the tide of things and the way things are changing due to modernization.

But you can adjust, by getting yourself modern forging tools like drop hammers and power hammers. It reduces the stress and fatigue that comes from using the crude hammer.

- Automating your operations by using industrial robots
  This might look quite extra, but the use of automated like industrial robots is a top way of advancing your business and they're not as expensive as you think. They can be helpful during the forging process by doing simple tasks like pick and place, metal cutting, and drilling, hence enabling you to focus on other areas of forging like heating, hitting, and shaping. They make the process easier and faster.

- Join online communities
  Joining an online community that specializes in blacksmithing would be one of the most amazing

steps you can take towards modernizing and advancing your blacksmith business. That way, you'll be able to reach your target market easily.

You could also learn better and newer blacksmithing techniques from people who are as passionate about blacksmithing just as you are.

- Identify your target market and market your product
The market for blacksmithing is not large, but very much defined. The possible targets are historians, war reenactors, and weapon collectors.

These markets pay well, so you don't need to worry about your chances of making good money off your hard labor. However, you have to learn how to identify and target them. Once you've identified your target market, it is time to shoot your shots. Be open to adopting modern forms of marketing.

Social media is a viable option; you could use Facebook, YouTube and Instagram. It will boost your visibility, which is very important in

31

marketing. Make sure you are targeting the right audience on these platforms.

Other modern methods of marketing are creating a blog or vlog where you can show off your skills to prove your credibility. You could also run an online ad. What's important is that you're visible.

- Ensure your forge is well secured
  The forge is a very dangerous place due to the heavy and sharp-edged tools lying around. It is also susceptible to fire, which makes it even more dangerous. You need to get your workshop inspected by a local inspector before you commence your blacksmithing business to prevent accidents.

Summary

Indeed, blacksmithing isn't as complex as it seems. It only requires a level of practice, patience, passion, and diligence to attain perfection. Yes! It is possible to be an excellent blacksmith and attain perfection, just like in every other skill.

Also, we have established the fact that blacksmithing has not gone into extinction. We still have blacksmiths who are doing fine today. You could connect with them on social media. This should be a strong motivation for you as a beginner blacksmith to push on with your dreams of being a blacksmith.

Finally, the only way to have a strong ground in the metal production industry as a blacksmith is to modernize your blacksmith business. To be on top of your game, you have to use the right modern instruments, connect with the right people, and shoot your shot aright. This would definitely cost you some money, but you can never compare the resultant advantage to the money you are spending.

You also need the right knowledge to excel as a blacksmith. This is why we have dedicated this chapter to educate you on all you need to know, even as you commence your journey.

See you in the next chapter!

Chapter 2

Basic Terminologies In Blacksmithing

Like every other craft, there are basic terminologies that define blacksmithing and are used consistently in practice. Because the art of blacksmithing is ancient, most of the terms and terminologies are also old and medieval. Hence, before you start pounding the steel, you need to learn these terminologies and their definitions.

Understanding these terms and terminologies would help you understand blacksmithing properly and all it entails. In these terms lies the real flavor of the art and you'll definitely need to hold on to them as you commence your journey.

There are so many terms we have in blacksmithing, complex, and simple terms alike. For simplicity's sake, we will be looking at just a few basic terminologies to help you have a great start in your blacksmithing journey. Below is a list of basic terminologies used in the steel pounding world.

Basic Terms In Blacksmithing

Forging: This has been defined extensively in chapter one. It is basically the art of heating, hitting, and hammering raw metal into shape using compressive forces. These forces include blazing heat, hammers, a die, and others.

Forge: A forge is a hearth used for burning coal at a very high temperature to heat the metal till it becomes easier to shape into the desired form. In other words, a forge can be defined as a stone-lined furnace where metals are well heated before taken to the anvil or die for hammering. In a case whereby the forge is indoors or in an enclosed place, a small vent will have to be set up in the ceiling to guide the smoke and ash away from the room. The vent will help to ensure the room is always conducive and comfortable, especially when the forge is in use. Every forge has a blower that helps to blow air from underneath onto hot coals. A forge could also be referred to as a place where the hearth or furnace is located. However, there are different types of forge.

Forging press: Forging press, also called press forging, is the process of shaping metal on a forge press. A forge press is a machine that shapes metal into a three-dimensional shape by applying gradual hydraulic pressure. This is a slow and gradual process. There are two press forging methods, open and closed die forging process. After the metal is placed in a die, the pressure is applied until there are slags of metal from the process filling the die cavities. By this time, the shaping is done.

Die: This is a flat-shaped tool that is used to form metal into specific shapes as the workpiece is hammered against it. The die could come in shape or plain most times. There are different shapes of die for forming metal into various forms.

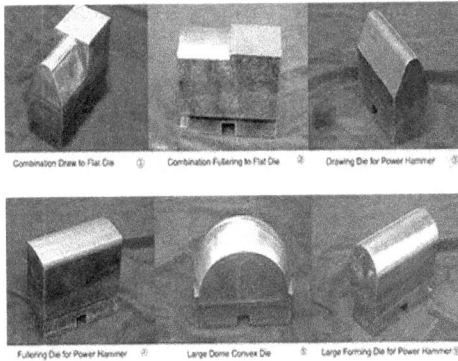

Combination Draw to Flat Die ①   Combination Fullering to Flat Die ②   Drawing Die for Power Hammer ③

Fullering Die for Power Hammer ④   Large Dome Convex Die ⑤   Large Forming Die for Power Hammer ⑥

Different shapes of die
36

Welding: This is the heating of two or more pieces of metals together until they become one. The forge is fanned up till the coals are burning at a very high temperature, and the metals are well heated to melting point. It is at that point when the metals are soft and malleable that they can become easily joined together. Even during repairs and rejoining of metals, you have to get all the metal pieces heated to a very high temperature. Welding is only possible when the metals are well heated, as fusion is only made possible by heat. After heating them, the smith could hammer the metals against each other, just like in the picture below. Hammering is very necessary when the metals are hard.

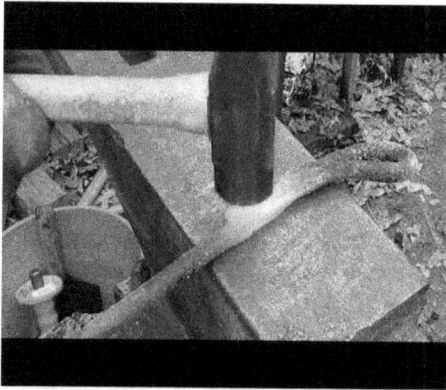

Welding of two hard metals together by hammering

Scales: The oxidation process where the air comes in contact with steel during heating is known as scales. The oxidation causes a layer of oxides to form a black coating on the steel as it is heated. To prevent such build-ups, the blacksmith needs to ensure that the steel is well positioned and the air blast is controlled. However, hammering the steel gently helps to break off the scales. Sometimes, scales can be inevitable, but if it becomes too much, the finished piece becomes less attractive.

Casting: This is the process of pouring hot, molten, and liquefied metal into a mold. The mold could be of any shape. The mold material is removed after the metal is cold and solid. The most common metals that are cast are aluminum and iron. The molds are usually referred to as castings.

The casting of molten metal in a crucible

Foundry: Foundry refers to the place or factory where metal is molten and cast.

Cold forging: This is one process of forging that is done at room temperature, without any form of heating. However, the metal is configured into the desired shape using dies. This is a much more comfortable and less stressful process of metal shaping. The energy needed to heat up the workpiece is well conserved. It has also been proven to be more economical as the end result is always a more substantial and higher quality product.

Slag: This is also called dross or tailings and can be defined as unwanted byproducts of smelting ore. This slag is gathered during the heating process. It is an excellent raw material for cement production and insulating material.

Smelting: This is the process of extracting metal from its ore through heating and melting. After the metal has been removed, what is left is slag and the eliminated gasses.

Wire pulling: This is the process of making wire by pulling metal through a wire-making die or hole.

Cold working: Cold working is the process of strengthening metal by changing its shape and causing a permanent plastic deformation known as "Work hardening." The increase in strength is a result of the change in the metal's crystalline structure.

Fuller: A longitudinal groove down the blade's length and on the two flat sides of the blade. A blade that has the longitudinal groove is said to be fullered. This is majorly down on weapons like swords to prevent them from whipping roughly and dangerously during swings. Fullering makes the blade of the sword lighter and more robust. It is similar to the small hammerhead. If made smaller and placed on the usual handle, it can make a good alternative for a hammer. Sometimes, using a fuller in place of a hammer can enhance the workpiece to make quicker progress as the fuller's rounded head sinks well into the metal than the flat-headed hammer.

Bolster: This is a part of a knife where the blade meets the handle. This junction is usually metallic or wooden. The bolster makes the knife firm, healthier, more durable, enables easy handling, and more convenient to use. Bolster in blacksmithing is also used to refer to a flat metal with punched holes. It is also called a bolster plate.

Ductility: This describes the property of a metal that determines the extent to which a metal can be stretched without rupturing. In other words, it is a property that measures the ability of a metal to withstand tensile stress. An example of tensile stress is the tug of war. The rope is pulled at two ends, and tensile stress is applied to it. Metals like copper are proven to have high ductility.

Buckling: Buckling refers to any form of a bend, wrap, buckle, or bulge in a metal.

Burr: When shaping a workpiece, there might be a need to cut off some piece of metal, especially along the edges. Sometimes after such cutting, the edge of the workpiece becomes rough. That roughness is what is referred to as burr.

Carbon steel: This is a special kind of steel that is very high in carbon. This type of carbon is used for only projects that require hardened steel as it can quickly be hardened by heat treatment. It is used for making things like bridges, cars, freezers, machines, axles, and every other thing that requires strong steel. Also, it has low ductility and zero flexibility.

Annealing: The process of heating up a metal to make it softer, so it can be easily worked on. To achieve this, the material is heated above recrystallization temperature for a suitable amount of time. The heated metal is left to cool slowly in still air or quenching in water, after which it is ready for further work (shaping or stamping). In the end, the metal is softer, ductile, malleable and easily bendable. Annealing aims to reduce the hardness of the metal and increase its ductility.

The process of annealing a metal

Crucible: This is a ceramic or metal container used in ancient times for melting metals. Crucibles are made with materials that can withstand very high temperature. The crucible must be made of very strong materials that doesn't stand a chance to be affected by the heat coming from the melting metal. Clay containers were the most popular type of crucible and it is still relevant today, as they are the most durable and perfect option for holding hot molten.

A clay crucible

Arbor press: Arbor press is a light pressing machine that is used to press tapered arbors to be machined on a lathe. It is a small hand-operated machine used to perform small jobs like configuring, installing, bearings and installing amongst others. However, it is useful in the blacksmith's shop for producing thing leaves for design and other pressed shapes for design.

Drawing out: Depending on the type of project, after forging, the next step is drawing out. Drawing out means to extend the length of a piece of metal, to make it wider, longer, or thinner. This is done by heating and hammering.

Flux: It is a cleaning compound used in fluxing the metal before welding to enable easy soldering. Some people use sand in place of this. Flux is used to keep oxygen from the surface of a metal to prevent oxidation, affecting the appearance of the final work and hindering the metals from joining properly. Most smiths used sands many years ago. Few years ago, people started to use borax and charcoal. Today inert gases are more common.

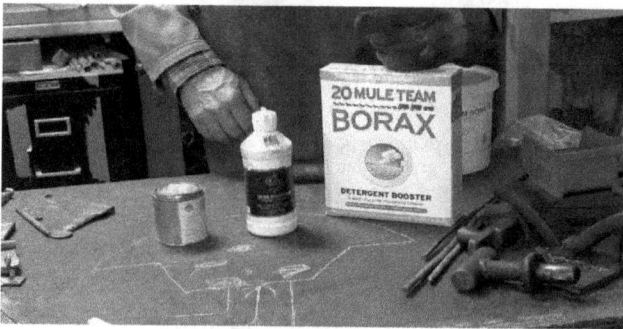

Borax flux

Scarf: Before two metals or more can be welded together, the ends of their metal must be prepared for fire. That end that has been prepared is called scarf.

Clinker: The impurities from the fuel at the bottom of the fire. They gather after every forging process and if it is not removed, it will block the air draught and reduce the heat. It can also stick to the metal and damage the blacksmithing process. It is very important to always remove clinkers when you are done working at the forge.

Chamfer: Chamfer means to remove the sharp edge of a metal. The vacant spot after removing sharp edges are also called Chamfer.

Pig iron: Pig irons are crude irons from a high blasting furnace. This iron is most times in rounded oblong bricks; an image that resembles a pig.

Pritchel hole: A round hole that is on top of the anvil so that you can push holes through the metal. It is near the hardy hole and used in combination with a pritchel (a kind of hammer with a small, rounded tip). Together, they can help make holes in the workpiece.

Hardy hole: A square hole on top of the anvil that is used to insert tools. It is designed specifically for tooling.

An anvil showing the pritchel hole and the handy hole

Cross-section: A view of something as if it had been cut through with knife.

Hardening: This is the process of making metals harder by tempering, quenching, or using chemicals to change the chemical make up of the metal.

Quench: This is a blacksmithing term that refers to the cooling of the metal by putting it in a cold water or oil. This will change the properties of the iron and harden it.

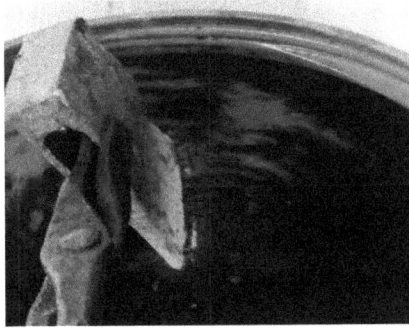

Quenching a heated metal in water

Temper: Temper is a heat treatment used to eliminate every form of brittleness as a result of the hardness. After the metal has been hardened, it is returned to the fire and carefully tempered to remove the brittles that formed during the hardening process.

Tang: The pointed part of a tool unto which a blade is lifted.

Anglesmith: Someone who welds metal into angular shapes.

Brightsmith: A person who works with bright metals like tin, copper, brass.

Farrier: A blacksmith who specializes in shoeing horses and oxen.

Summary

These terms and terminologies consist of names of tools and basic processes involved in blacksmithing, thus enhancing you with a balanced knowledge of what this craft is all about. Endeavor to hold on to these terms; the knowledge will enable you to gain a perfect understanding of the more in-depth things that will be exposed in the rest of this book.

Some of these terms will be explicitly discussed in the coming chapters.

Chapter 3

Blacksmithing Tips And Tricks

These are basic and simple tricks you could practice as a beginner blacksmith. These tricks are fundamental and can help you gain mastery over the craft more quickly than you ever thought. Some of them are quite technical and you will need to practice several times. Hence, try to pay close attention to these tips and tricks that will be listed here. Don't be in a hurry to get through.

A good number of the tricks mentioned here are gotten from great and professional smiths who have been in the field for several years. Many of these tips have been around for donkey years, right from the eighteenth century when blacksmithing started.

Studying them is just like reaching out into the ancient world and trying to familiarize how it worked with them, how they were able to prune this skill and produce the finest of art, and still pass this art down to the younger generation as a heritage. For me, that is deeply fascinating.

However, in as much as we will want to reach the borders of the earth and breathe in the air of inspiration that the ancient walked in, we also have to accept the fact that not all of these old and conk skills are practicable in our time.

Yeah, from our study of history, we realized that the metal production world has advanced greatly and the zone of blacksmithing has been adversely affected. Hence, new methods of doing blacksmithing have been developed; Very much easier and faster methods that won't only help you become a great blacksmith but also present the craft in an easier light. Simple tricks have replaced crude and tiring methods. Sincerely, we owe technology many thanks for that. Blacksmiths of old would have given anything to have the resources we have today!

Now, the once conk and reserved craft has been made plain for the simplest person on earth, thanks to technology. With these tricks, anyone can become a blacksmith, yes anyone!

However, in simplifying the craft, some of the tips that accentuated the uniqueness of the craft of the ancient smiths have been put aside. Well, what did they say again about trying to eat your cake and have it?

The crudeness of the craft was an inspiration behind the spectacular artworks they created. If what you seek is to create unique and timeless products like of old, you can't possibly do that by simply following these tricks. That does not mean you can't create something still universally unique and distinct. In fact, some contemporary blacksmiths are doing very well today and creating history in blacksmithing. We will be connecting with the tricks and tips they use in the projects.

I have sorted out the kinks from the details and have come up with a list of tips that can help you achieve your blacksmithing goal.

Let's see how we can join the tips from ancient smiths in addition to the modern tricks and discover ways of making our own timeless products extraordinaire.

Basic Tips and Tricks of Blacksmithing

In chapter one, we mentioned that there are four basic blacksmithing processes: Heating, holding, hitting and shaping. These processes were explained to some extent, but we will be breaking them down and looking at some little tricks involved in each of these processes.

We won't go on to define them; we will only break down these processes and analyze important tips under each of them.

Heating

Blacksmithing is mainly about heating, reheating and heating, even while other processes are on. Sometimes, after hitting and hammering, you would still need to heat up your metal to get the perfect touch of what you want. This is why your fire must never go out; you need the right amount of temperature handy.

There are basically two types of forge; the conventional and the contemporary. The conventional forge is the use of coals, firewood and a blast furnace.

Contemporary forge is of various forms; some of them are electrical and could help you regulate the heat and get your project done faster. Again, you don't need to keep trying to steady your fire and keep it on. This forge is a stress reliever. An example of contemporary forge is induction forging. More will be said about this form of forging in the next chapter.

Most of these forges are quite expensive, especially the induction forge. Talk about the cost of relieving stress. Well, don't begin to feel sad because you can't afford this kind of forge. All of the different forges work the same way; the only major difference is that one makes it easier by controlling the heat temperature. However, if you can't afford it, don't make it kill your vibes to smith. The early smiths made the best of instruments and hand tools with their coal forge at a time when there was no help from technology. Forges are the only instruments used to forge and the type of forge used in making a project doesn't accentuate or reduce anything. The major benefit of using a contemporary forge is the ease and nothing more. It makes it easy for you to regulate the heat and maintain a tidy environment. If you don't know how to smith well and you get an expensive outstanding forge, you will make the most horrible project of all time. So, getting the tool is not as basic as sitting down to learn the right tricks and tips.

You just need to do what works for you with what is available to you. Besides, you are only a beginner. Don't let the pressure of getting things right get to you now. It is way too early. Having pointed out the importance of steady heat in the forging process, there are two other things that stand out; temperature and color when forging.

Temperature: Temperature here is the measure of heat treatment given to the workpiece you are forging to get the desired result. This is a piece of information you need to have handy and at the tips of your fingers. It will help you to get through the forging process quickly. The highest temperature in forging is the melting temperature. Every metal melts at a different temperature. Steel melts at 1500 Celsius; aluminum melts at 660 Celsius. You can set your forge temperature checker if you are using an electronic forge. However, if you don't have a means of checking the degree of temperature, you need to master the colors of steel in forging.

Color: Some practicing smiths still make mistakes today because of their ignorance of this. Knowing the colors of your heating metal and what it indicates in blacksmithing will save you a whole lot as you start your smith career. It will also give you an edge of advantage as you will escape some mistakes and falls that common blacksmiths make.

Now, what does color represent in forging?

During the heating process, some colors point out the temperature and usability of the metal.

If you cannot measure the temperature in degrees, you should be able to tell the colors. If you are using a local forge like coal, you cannot tell the temperature in degrees. Hence this information is very candid for you.

At temperature 400-500 Celsius, metals like lead and alkalis turns silver.

When iron is heated to high temperatures, it glows bright red then changes to orange. If you leave it for a longer time, it heats up and changes from orange to yellow. From yellow, it changes to color white and stays at white. Once you leave it after that time, the metal starts to melt and if that's not what you want, you are going to have a mess of molten metal.

Increase your forge's heat temperature, but stay around to ensure that your metal gets heated to the right level it should and it doesn't go beyond.

This is why you need to do your forging under bright light. If you are outdoors, cast your shadows over the forge to help you get a perfect discretion of your metal color.

Picture of a glowing metal

Hitting and Holding

There are a few tips under here. The very first is hammer control.

You need to learn effective hammer control. The wrong hit could mess your work badly. Also, you don't have to hit the metal too hard to give it shape. The right volume of hits in the right place can give you the result you require. Effective hammering is not always about the degree of strength and pressure. Hammering is majorly about striking the metal with the RIGHT degree of pressure at the right place. Asides the wrong impact bad hammering can have on your project, it could also lead to accidents like hitting the hammer on your hand or dropping the hammer to your feet.

A good hammer control will save you so many accidental injuries and strains. Be precise and calculative before you start to strike your hits. Take out time to practice this until you get it right. When you have a metal of great value to work on, you will appreciate practice efforts.

Picture of a person hammering a heated metal

Also, try to build smart and sharp hand-eye coordination: When you are forging metal, you need good eye-hand coordination. Your eyes must be focused on what your hands are doing, even as you try to put a lot of things in place simultaneously. Blacksmithing will require all of your attention. You can't afford to be bodily present and absent-minded. If you are distracted by problems happening in your life, take some moments and cool off before you try to take the heat of the forge. Your whole mind and body have to be concentrated on whatever you are working on. It will help you maximize time and reduce the usage of fuel. Full concentration also prevents errors and external injuries.

Endeavor to use only the right tools: Never use something as a makeover for a poor or bad tool. Try to ensure you have double of every tool. If not possible to get multiple, keep the ones you have gathered well. I'll talk about tool maintenance and usage in the next chapter. Three basic tools necessary for blacksmithing are hammer, anvil and tongs.

Shaping

There are different steps in shaping. This is a creativity zone and requires much patience and unique technicalities for varying projects. However, everyone uses some basic tricks no matter the design you want to achieve or the piece you are working on. We'll be looking at bending and fire-wielding, amongst others.

Bending: Bends are much easier at a highly heated temperature. The best time to bend and shape your metal is at its melting point. At that point, it is easy to draw out, to create whatever form of curve you desire and shape bend the metal. There are various ways of bending; however, when you are bending with your hammer, be careful to keep to the marked place, so you won't have to keep doing trials and errors. Make sure to hammer gently to avoid error. Heat the part of the metal you want to bend to the hottest temperature to make it easier and faster for you.

Fire welding: The process of joining metals together through fire welding is top skill. Ensure to heat up your metals to a melting point, then hammer it together. If you are heating two different metals (light and hard metals), make sure to heat the hard metal early and well before heating the light metal. For example, you want to heat copper and iron together, you have to first start heating the iron minutes before you put the copper in the fire. Once both of them are at melting point, you bring it out at the same time and start hammering. In no time, you will be done.

Below is a list of random tips you need to be familiar with and are not directly related to any process.

However, it is mainly for coal forging. They are not less important than any of the tips we have mentioned above.

- Allow your metal to heat well.

Resist the urge to bring out your workpiece from the fire immediately it turns cherry red. You might feel your metal will melt when you leave it for long, but that is not true. Metal doesn't melt that easily. However, with light metals like copper and bronze, you have to be more conscious of the heating time. A well-heated metal makes your other work easy. If your metal is poorly heated, it will take you a long time and more energy to hit and shape it.

The only reason you should bring out your metal early is if you are looking for a hammered look, but you would get frustrated if you try shaping it at that point. It is just like you never put it on the fire. Fire is one big deal many of us don't know how to deal with. How much fuel or coal does your fire need, how long does it take to get to a certain temperature, what side of your forge is the hottest? Is the air too fast and interrupting the fire? These are things you will practically learn as you start forging.

- Strike when it is hot

You have to be conscious of time when shaping your workpiece. The best time to do all your shaping is when the metal is red hot. Your tools easily penetrate when your metal is hot and allows you to have a perfect strike as the metal is softer.

If you fail to strike at that time, the initial efforts you have made in heating the metal becomes a total waste as you commence the struggle of trying to shape a hardened metal or trying to heat it again. Now, you understand where the old adage, 'strike when it is hot' was gotten from.

- Make sure you are using the right metal for the right project

Some projects require hard metals, while some require light metals. If you use hard metals for a light project, you will spend a lot of time trying to melt and soften it. The same thing is likely to happen if you use a light metal for a hard product; you could spend the whole day trying to harden it.

Copper and bronze are light metals and cannot be hardened by heat. Their carbon content is very high. The only way to harden metals with high carbon contents of about 53%, just like copper, is by cold working. If you are working on something flexible, copper is a viable option. It is very malleable and easy to bend. There are also some types of iron you shouldn't try forging.

Do not ever try forging a cast iron. Any metal with its carbon content at 2% is called a cast iron and is a bad choice for a blacksmithing project.

- Tend to your fire

Even after you take out the metal from the fire, don't be quick to put your fire out. Neither should you ignore it. The fire should be on throughout the process of your forging until you're done, lest you might have to keep putting on the fire. Heat is needed throughout the state of blacksmithing, just as we mentioned before now. It is the most important ingredient in blacksmithing. You cannot afford to let your fire die and lose its heat.

- Keep your tools handy

Make sure your tools are very accessible at all times. You should have all of them in a stationary place. In case of emergencies, you don't go tearing your workshop apart because you are searching for a particular tool. Sometimes, in searching for your tools, the heated metal you removed from the fire would have started getting cold and thus, constituting a waste of time. Time is a precious asset in blacksmithing.

- Quench slowly

Don't be in a hurry to get your metal cold and throw it into a bucket of water. It could completely shatter your work. Cooling your workpiece too quickly will cause it to weaken and brittle. You wouldn't want to lose your work for whatever reason, so quench very slowly and don't be in a hurry dump your metal into a bucket of water or oil.

- Empty that ash trap

Kill and deal with the habit of keeping a load and clogged ash trap. Asides from that, it doesn't make your forge appear neat; it could cause sparks to fly everywhere while you are forging a new project and likely pop on you. A clogged ash trap could also cause clinkers to rise from the fire and roughen your metal.

Result of a clogged ash trap

- Brushfire scale from your metal

During oxidation, when the temperature of your metal is changing, there is this black coating that tends to firm on the metal. If ignored, it would make your metal rough and pit the surface. Brush it off immediately. Use a wielding brush to get rid of it.

Picture of a wielding brush

The metal could get really rough and might require more effort to clean than it normally would. The best way of handling fire scales is to brush it away immediately; if you fail to do that, it could pop on you when you start hammering and give you terrible burns.

Nonetheless, there is one tip that has not been mentioned so far, some smiths mostly neglect it and they end up getting frustrated at the end.

If you don't get any other thing, get this: BE COMFORTABLE MAKING MISTAKES.

It sounds odd and meaningless, right? Who gets comfortable with mistakes? There is no way someone can get comfortable making mistakes; how in the world?

Well, for all your questions, this is a top smithy code and rule. You are only a beginner. Don't be too hard on yourself for making a mistake. You might have to make several mistakes before you get it right. No professional smith was born professional; they had to go through the route of making mistakes and trying it again until it was a success.

Have an open mind towards mistakes; it will help you become better. Have fun experimenting with techniques and new styles.

Be open to learning even after you have perfected a number of projects. Read books, watch youtube videos, visit blacksmithing websites, and join a community of professional writers. It'll help you on your journey.

You could discover your own trick and flow in the spirit of the blacksmith.

Summary

You would notice that the last tips focused majorly on shaping. Shaping is a very huge task in blacksmithing. Indeed, it is the climax of the heating, hitting and hammerings you've been doing. Hence, you have to just get it right or ruin your efforts in the previous stages.

All of these tips are very valid and practically proven to be helpful. It will save you a whole lot of stress and energy that could be wasted by ignorance.

Endeavor to practice these tricks over and over again before you attempt to make any products. If you don't get it right at the first trial, don't let it get to you. You could always keep trying until you attain perfection.

Try it out patiently, get comfortable making mistakes, assess yourself clearly and try again.

A Short message from the Author:

Hey, I hope you are enjoying the book? I would love to hear your thoughts!

Many readers do not know how hard reviews are to come by and how much they help an author.

Customer Reviews

⭐⭐⭐⭐⭐ 2
5.0 out of 5 stars ▾

5 star		100%
4 star		0%
3 star		0%
2 star		0%
1 star		0%

See all verified purchase reviews ›

Share your thoughts with other customers

Write a customer review

I would be incredibly grateful if you could take just 60 seconds to write a short review on Amazon, even if it is a few sentences!

>> Click here to leave a quick review

Thanks for the time taken to share your thoughts!

Chapter 4

Getting Started with Blacksmithing

Phew! Are you ready for this right now? I mean, right now? I will like to imagine you putting on your gloves and rubbing your palms against each other as you get prepared to be launched into the world of blacksmithing.

We have gone through several hitches and pointed out basic landmarks on this journey for you. None of them is less important, so don't look down on any.

Yeah, we are just getting started. There are still several basics we need to trash out before you get on track to get your first blacksmith project in the works. If you're feeling what I'm feeling as I am typing this right now, you'll be bubbling with excitement. I really can't wait to see you set off.

It feels so close to the end of the track when it is only the beginning. Isn't that amusing? Well, as much as I'll love to go on and on bubbling with excitement, let's get to the teeth of this project.

Let's get started.

Basic Tools And Supplies You Will Need

Again, we will be looking at the basic tools and supplies you would need to get as you set out, but this time from a more detailed outlook. There are a few things that are not necessary for you to get at first. They are majorly for professional projects. You don't need such tools now. The list below comprises of very important tools that you might not be able to work without.

Steels

This is the first tool you will be needing. You need to get the right steel for your project. There are many types of steel, but as we have mentioned in the previous chapter, you need to select the right one for your project, as there are specific tools for each project.

- Carbon steels: Studies show that this is the most popular steel found amongst blacksmiths today because of its numerous amazing properties. The level of carbon content determines its strength. The lower the carbon, the lesser the strength. There are three small categories of its level of carbon content. Low carbon steels, which are normally mild, is up to 0.3%. These steels have low cost, high formidability and low strength, such as wire, nail, chain, etc.

Medium carbon is measured at (0.3-0.6% carbon). This is much better than the former. It has relatively good strength, toughness, and ductility. Elements are screws, axles, cylinders, etc.

High carbon steels, which are relatively hard, is more than 0.6% carbon. These are characterized by high strength, hardness, and ductility. Its element are heavy metals like hammer, screwdrivers, rolling mills, etc.

- Alloy steels: This steel contains alloying properties and elements. Its properties are popularly characterized by hardness, strength, formidability, ductility and corrosion resistance. It is a great choice for blacksmithing, especially as a starter. It is very much easy to handle and requires less effort during formidability. Some of its elements are very shiny and attractive. Examples are aluminum, silicon, titanium, nickel, manganese, chromium and copper. Copper is one of the most commonly used alloy steel and a good choice to consider for practice as a beginner.

- Stainless steels: These are alloys that contain 10-20% chromium. At 11% chromium, the steel becomes super resistant to corrosion. This steel doesn't change its color or react chemically to air or water. It is used in making many household tools and utensils like knives, spoons, plates, cups, bowls and others.
- Tool steels: A steel of great composition of various quantity of chemicals; cobalt, molybdenum, and vanadium. It is highly durable and has a strong heat resistance property.

Forge

This is where your blacksmithing process begins, at the forge (it is also called a hearth). It is very necessary for heating, annealing and hardening amongst others. It is the next important item to get in your purchase of blacksmithing tools. You need to choose the forge that is most affordable and accessible to you. You want something that you can use for a long time and has a good heat temperature control. There are several of them; some are conventional and others contemporary, and we have a brief expository description of each below.

- Coke/charcoal forge: This conventional forge was used by the early smiths around the 18th century. Then, it was majorly charcoal and bituminous coal that was used by the smiths. Some smiths also used small wood to make fire. Today it has been modernized and industrial coke has been introduced by technology. However, the design still remains the same irrespective of whether it is coke, coal, or charcoal. This type of forge is also called a hearth or fireplace.

The only means controlling this fire is by the amount of air, and the volume of fuel. If you desire a higher heat temperature, you could increase the heat by adding fuel and allowing more air flow. Although this forge has lasted thousands of years, it still has the following features:

Tuyere: A channel, usually a pipe, through which air is forced into the forge.

Hearth: This is above the tuyere opening and is where the fire is made, where the burning fire is.

Bellows or blower: The entry of the tuyere, a means of forcing air into the air channel.

The smith controls the heat temperature to the kind of work he is doing by adjusting the fuel and the airflow. He could also adjust the length and width of the fire of such a forge to accommodate different shapes of work. If the smith is using coal and coke in one fire, the coke should be placed in the middle as it easily and quickly catches temperature and is much easy to set on fire. The coal should be placed just round about it.

There are a variety of this one kind of forge, with different designs, shapes and color. However, all we have listed above remains the basic way that a coal or coke forge can be used, irrespective of the shape or color.

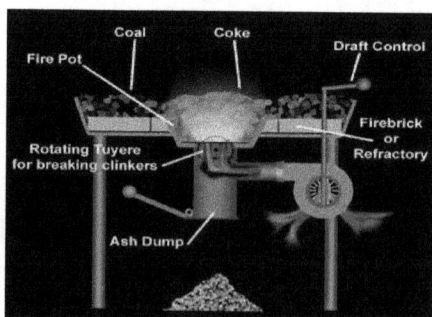

- Gas forge: Natural gas is filled into a cylinder, which is connected to the body of the forge. The gas cylinder is usually ceramic. The forge is a burner chamber lining. To increase the heat treatment, heighten the mechanical blower by taking advantage of the Venturi effect. It varies in size, large forge with big burners. There are also small ones that use simple and propane torch; it could be carved out of firebrick. The forge gas is very easy to use, a great choice for a beginner, eliminates the stress of increasing temperature, set coals on fire and all of that. The workplace is clean and the fire is consistent. Unlike with the coal forge, where you have to be on standby to ensure the fire doesn't go higher than you want or it doesn't go out. However, it cannot be used to forge very heavy metals. Asides from this limitation, it is a great choice for any task.

- Solar forge: This forge uses sun power to melt materials with low melting points. It cannot afford as much heat temperature as other forges. This device can only work feasibly well in the desert or places where there is a lot of sunlight. It would be a waste of time trying to use this in a place with poor sunlight. The device includes a large rectangular Fresnel lens and is positioned on a hollow standing cylinder like a lid. The material is placed inside the cylinder, where the rays of sunlight and its radiation are channeled into through the lens. One advantage of this forge is that it is easy to build and consists of basically two parts. You would need to use a mask and fireproof clothing as the lens's heat is so much and can be harmful.

- Hydraulic press forge: A press forge is a second group of forging machines regularly used in impressive forging. The hydraulic press forge is slow-moving compared with the mechanical press forge. Pressure is applied to the top of the piston. This forge is great for forging aluminum, alloy, and magnesium into wheel and bicycle components.

- Drop forge: The workpiece is created by hammering a piece of hot metal into an approximately shaped die. This is pounding or shaping a heated metal between two dies with a drop hammer or a press. One of the dies is usually fixed and the other is acting by gravity. The device is majorly used for making large forgings.

- Finery forge: This is a water-powered mill used for producing wrought iron from pig iron. This process is called decarburization in a fining process. The cast iron is liquefied by oxidation. By this process, carbon is removed from the molten cast iron. It is an old method of forging

- Induction forge: Induction forge is an electrical furnace in which heat is applied to the metal piece by induction to make it malleable. This forge heats metals to between 1,100 and 1,200 Celsius. This is a clean, energy-efficient and well controllable melting process compared to the coal forge. However, it is limited by electricity. If you can manage the limitation, it is a great option for you to consider as a beginner. Many modern smiths use this foundry.

Hammers

This is one tool you can almost never do without. You need it through all of the processes, especially the hitting and shaping. There are different types of hammers that blacksmiths use and as you will guess right, they are used for unique projects. As we said, endeavor to get the right tool for the right project; choose the right hammer, not the available one. However, most of them are multi-purposed, so you don't need to get all before you can commence. Two multipurpose and a number of specific hammers.

Below is a short list;

- Chasing hammer: Tthis hammer has a springy handle and is fat on the down head but thin where it is attached to the head. It increases hammer control and reduces fatigue. One side of the head is very large, about *3 of the other side. Both sides are round and precise. Just like the name implies, the large head is used to strike the head of chasing tools (they will be listed below) and punches; it is not to directly contact your work.
- Cross Peen hammer: This hammer's head is flat and pointed on one side, while it is round on the other. It is a multipurpose hammer. It could be used for shaping, riveting, and striking steel tools as well as forging.
- Creasing/bordering hammer: Sometimes, it is used before the raising hammer to create radical crimps and creases in the metal disc. The bordering hammer is also used to create a rim on plates and other metals with its rectangular head. The cross-sections have a well-pronounced vertical curve. The usual weight is 200-300 grams.
- Raising hammers: Raising hammers are used on the metal's outer surface to form it into bowls,

valleys, and hollow forms in conjunction with a raising stake. It is used to create hollow metals. The faces of the head are rectangular with a slight curve. The weight and size are varying.

- Riveting hammers: A useful general-purpose hammer especially for spreading heads of rivets. It is wedge-shaped with one side sharply vertical and the other side is either round or square is used to refine rivet heads.
- Ball peen hammer: A one rounded peen faces useful for light forging, spread rivet heads and striking steel tools. It is the most recognizable style of hammer outside the field of metalsmithing. One of the sides of the head is cloned and the other is largely round.
- Wood mallet: Wedge-shaped wood mallets can be used for a large variety of tasks when you file, shape and sand the hard word, which is the materials used to make this kind of mallet. In a case whereby you are in urgent need for a riveting hammer or peen hammer and you can't find any, you could shape this wood head into the desired shape. The round-shaped head is used to form crimps and a great choice for

delicate products; it doesn't mark your work's surface.

- Brass mallet: With less reverberation, this hammer is used to prevent your metal from being marked during striking steels. It also prevents unwanted movements and guarantees a precise and good landing when used with stamping tools.

- Rawhide mallet: With a wide and heavy wood head, it is very similar to a wood mallet and will not mark your metal. The effect is heavier and more durable than that of the wood mallet. It can be gotten at different face diameters and good for increased driving force.

- Dead blow mallet: This hammer comes in a plastic or rubber head, which reduces shock and prevents the face from being scratched. The two ends of the hammerhead are round and it is quite heavy as the inside is filled with steel shot. When in use, the shots move from one end to the other, stabilizing the hammer, reducing reverberation, and stabilizing it. It increases the driving force as the hammer comes in contact with the metal.

- Plastic mallet: A unique metal mallet that could also be covered in plastic to prevent marks from covering your work surface. It usually comes in plastic covering, which is removable faces. The plastic material is high in density and non-porous nylon

- Goldsmith's hammer: With one peen face and one flat face, the design of the cross sections is very similar to that of a riveting hammer. This is a well-balanced lightweight hammer used for riveting and hammering.

- Planishing hammers: The cross-section of this hammer comes in two flat shapes; square and circle. As the name implies, it is used to refine curved and flatforms' outer surfaces to remove hammer marks and polish the metal form to mirror finish.

- Forming hammers: This is used inside the metal to create hollow shapes, just like raising hammers are for the outside. It could be used to refine and create a curved surface. It is used with stakes and wood forms. The hammer's cross-section is cloned and should match the curve of the metal.

- Embossing hammer: The faces are usually smaller in diameter and although it is used as a forming hammer, the heads are in small diameter. They are used to create elevated areas by hitting the metal from behind to form a hollow.

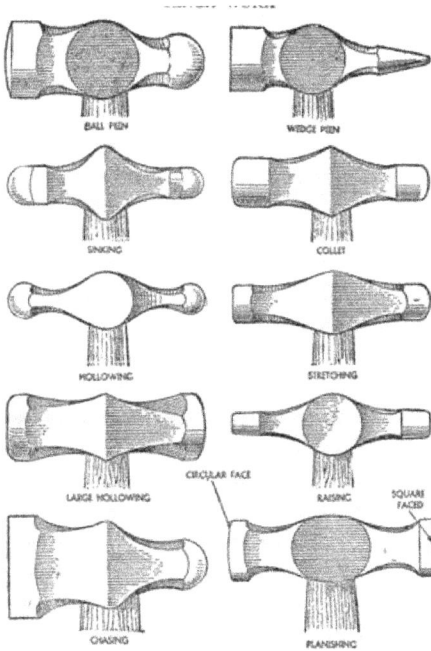

Picture of different hammerheads

Anvil

As explained in chapter one, the anvil is a heavy metal on which the heated steel is hammered and shaped.

This metal is usually rectangular and hollow. Anvils are normally massive but also come in varying sizes and weights.

There are different designs of an anvil; most of them look similar but are distinguished by salient features. These features are made to fit the project the smith is working on. There are anvils specially made for farriers, cutlers, armorers, coopers, and for general smiths.

The common blacksmith's anvil is made of forged wrought or cast iron. It has two faces, the top long and flat surface that is mostly used for forging and the base used as a stand. This base is mostly square in shape and smaller than the top face. Most anvils have two holes on its top surface; the pritchel and hardy hole. The pritchel hole is used to create holes in the metal piece using the pritchel or chisel; this process is well explained in chapter two. The hardy hole is used majorly for holding tools like chisels, tongs and pritchel. When purchasing an anvil, ensure to look out for one with a strong metal; avoid anvils made with inferior iron; it can easily break. A good anvil is supposed to create a rebound during hammering, i.e., to absorb the pressures from the hammer and make the process of hammering faster. An anvil without a rebound will cause you more stress and early fatigue.

The anvil is usually used with a hammer and anvil stand. The anvil is placed on the anvil stand and the hammer is used to hit the metal piece on the anvil.

A regular anvil

Anvil Stand

It is unwise to place your anvil on the floor; it would require that you bend to hit and shape your metal piece, resulting in waist and spinal pains. The anvil stand is made for you to position your anvil on. It also comes in different shapes, styles and heights. The most common shapes are circular, square and rectangular. The anvil stand is usually made of metal, wood, or a combination of both. It doesn't matter what it is made it; what is important is that it is strong enough to carry the anvil and give it a good balance. It comes in varying height; ensure to select a height that works well for you, one that is not too high or low and will give you an even balance to hit, hammer and shape.

Some anvil stands come with a chain. This chain is used to keep the anvil in perfect shape and prevent it from falling off. You could also get a chain to hold your anvil firmly on the stand. Most wooden anvil stands also come with hangers, where heavy instruments can be placed.

However it comes, just ensure it is steady, has a balanced height and matches the strength of your anvil.

Tongs

This is a hand tool used to pick heated metal from the forge or furnace. You definitely can't use your hand to do that. It is a second important tool after the hammer and anvil. It is also used to manipulate or reposition the workpiece with ease and accuracy during forging. Thirdly, it can be used as a holding tool to hold the metal in place and keep your hand away during hammering. A tong is usually long and straight. It comes in varying lengths and has different shapes of grips for different purposes. The tong's length depends on the kind of projects you do; the available range is 15 to 40 inches. Well, also keep in mind that the longer the tong, the weightier it becomes.

We will look at the types of tongs at this point. There are different types and shapes of tongs, customized for different applications.

a. V-bit tong: This is built with heavy-duty forged steel and can be used to hold a variety of stock. Highly versatile. Its easy weight and versatility make it a great option for beginners. This tong has a unique circular shape with a hollow at one end that meets at a cross-section for holding the metal in place. The other end after the cross-section is straight to help you maintain a strong and firm grip on the metal. It can hold a square and round metal equally. V-bit tongs are popularly used for knife making.

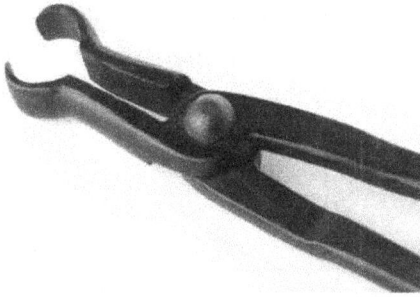

b. Wolf jaw tong: This is one of the most versatile tongs there is. It is also a good multipurpose tong and very strong to hold your workpiece in place during forging. It is relatively lightweight. The wolf jaw has a wavy tooth-like shape that allows it to be used on different shapes of metals (round and square). However, it is not as hollow as V-bit tongs and cannot be used to hold hollow metals.

c. Z Offset (Z-Jaw) tong: This is a perfect option for flat metals. It could also serve for all shapes as it has a unique Z shaped end that enables it to hold metals in a good place for cutting and forging. It is not a general-purpose tong and is used for making flat projects like knives and blades. It is not a normal beginner tong and is used for specific purposes alone.

d. Spreading tong: As the name implies, it is a specialized tong used for bending metals into peculiar and detailed shapes and majorly used in making spiral metal baskets and scrolls. It is not your usual tong.

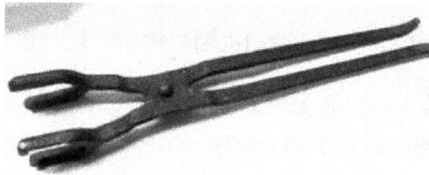

e. Hammer tong: This particular type of tong don't have general application. It is used in the forging of hammerheads and guarantees a good grip to help you keep the hammerhead in place and rotate it on an even axis when shaping it.

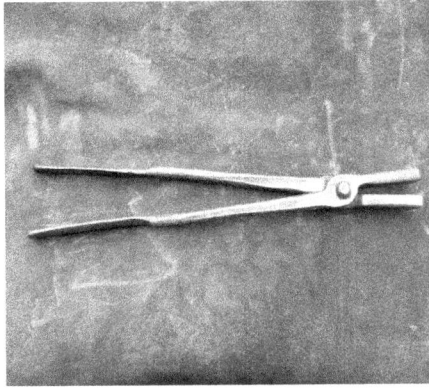

f. Rivet or pick-up tong: Shaped with a long straight end, the rivet tong is generally used to pick heated metals from the forge and manipulate it to ensure the even spread of heat. It cannot be used during hammering to hold the metal.

g. Bent knee tong: This is another hollow tong used for picking flat metals. It cannot also be used during hammering. It allows the blacksmith to easily pick a workpiece from as low as the forge floor without hurting your hands.

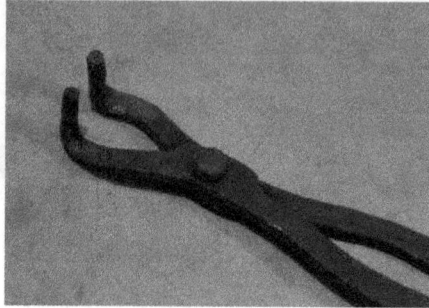

Having all of these tongs are nice and quite essential. If you don't have the funds to get them now, you could simply get a universal tong for versatility, like the V-bit tong and wolf jaw tong. They are popularly used amongst beginner blacksmiths for multipurpose functions.

Water Trough/ Quench Tub

A water trough and quench tub are the same tools, mostly bucket-like, and used to hold water for cooling the metal during hardening. You could make-shift one for yourself from a bowl or bucket. The water in the quench tub can also be used to reduce the flames from the forge when it becomes too large. It is very important that you have this close by, even if you don't intend to harden your metal. You might need it for your fire.

Chisel

Blacksmiths require chisels to cut both cold and hot metal. To cut cold metal, the chisels to be used are comparatively short and thick, whereas for cutting hot metal, they are thinner and longer. Chisels come in different shapes and sizes and are best made from steel with 0.8 percent carbon content. In the absence of such carbon content steel, motor-vehicle coil and leaf springs can be used as a fair substitute.

Often, tradesmen call upon blacksmiths to make chisels for them. These chisels must be hardened and tempered to suit the particular purposes for which it is being made.

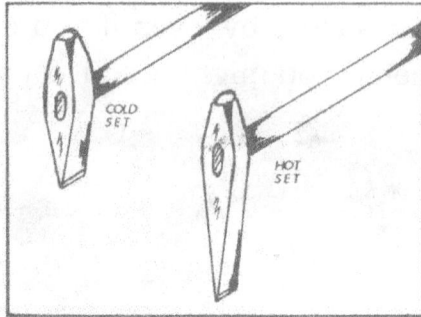

Vise

This is the conventional tool for holding a metal in place during hammering. It comes in strong metal that can sustain the pressure from the rebounding of the hammer. Just like the tong, there are different shapes for different purposes.

Some come with stands, some come with bolts that enable you to place them permanently on a bench. It is also used to hold the metal through other processes like cutting and bending.

Whether it is on the bench or standing on a stand, the vise must be strong enough to hold the metal and equally sturdy. The vise is a very important tool when it comes to bending, shaping and forging. The smith must get just the right vise.

Punches

This is a tool used to make holes in metals during forging. There are different shapes of punches for making holes. One uncommon punch is the decorative punch; it creates impressive holes of images in a metal.

Blacksmithing Safety Rules and Safety Equipment

Blacksmithing can be a very risky craft. Talk about flames from the fire, heat from the metal, heavy objects and fire sparks. It is not a funfair craft at all and should have some rules and restrictions to guide the smithy's conduct to prevent accidents. There are also safety equipment that must compulsorily be in place if the blacksmith will be guaranteed a smooth operation. Some states have officials in charge of instructing and advising the safety equipment you should have as a blacksmith. Below are some of the basic safety rules and equipment you need to know and have

Safety Rules

1. Use the right equipment for all projects: Using the wrong tools could be endangering to your safety. Don't hesitate to fix your tools once they get spoilt; this is another reason you are encouraged to have multiple tools, so you won't be stranded when one gets bad. Use a hammer when you need one and a chisel when necessary. Also, avoid the use of malfunctioning tools

2. Handle all tools with care: There is a safety rule that says not all dark metals are cold. Do not

assume that the metal has been long away from heat, so it should be cold. A metal could be dark and piping hot. Always be proactive and give room for peradventures as you work. Handle your tools also with care.

3. Don't play pranks in the forge: Jokes and play can be dangerous around heavy pieces of machinery; avoid playing pranks around the forge. The forge is not a place for fun. Prevent children from entering into your workplace, especially when you are working.

4. Ensure that your forge is well ventilated and lightened: Forging in a dark or poorly ventilated place is not ideal. You could hurt yourself in the dark or get suffocated by the smoke. Make sure the windows are up and there is sufficient light to enable you to see clearly.

5. Keep a first aid box within reach: Accidents are liable to happen even when you are most careful. Make sure a first aid box is very close to you.

6. Pay attention to your health's tone: Do not try forging if you don't feel up to it. Endeavor to rest

once you feel weakness in your body or fall short of health. Blacksmithing is not for the sick.

7. Turn off the fire once you are done forging: Yo not forget to turn off the fire when you are done. You can use the water in the quench tub to coal off the flames from the coal.

8. Have a specific place for keeping your tools: You could build a rack or shelf for placing your tools.

Safety Equipments

1. Hand gloves: The hand gloves prevent blisters from the roughness of the metal while blacksmithing. Some metals can be very rough and hard; you need to protect your hands from getting injured. Also, dealing with fire could be so tricky and you can mistakenly touch a hot object that will burn you—more reason why you need to guard your hands.

2. Apron: The apron helps to prevent sparks from jumping on your body or soiling your clothing. The apron should cover you from your chest

region to your knee. The best material for an apron is rubber.

3. Shades: The sparks from the fire could be blinding and you wouldn't want to lose your eyes to a craft, no matter how much you love it. Use google or eyeshades to protect your eyes. You could use some dark shades or large eye shields for welding.

4. Earplugs: Hammering could be quite loud. A set of earplugs would help you reduce the effect of the loud noise from hurting your ear drums.

5. Safety boots: This is necessary to protect your legs from coming in contact with dangerous tools or molten metal on the floor. Even though you are to sweep your forge floor every thirty minutes, you still can't be guaranteed that the floor is safe for your bare feet.

6. Shovel: Coal could jump out and land on the ground anytime; this is not something you really have control over. So, keep a shovel handy to shove it out immediately. This is very important

if you are indoors and working near combustibles.

7. Fire extinguishers: Not all fires could be put out with shovels or water. You need to have a fire extinguisher in case there is a fire breakout or accident.

Setting-up Your Blacksmith Workplace

A blacksmith's workplace is his office and should look like one, not a jumble of iron, coal and tools. Everything should be placed in order and the whole place should be tidy. Starting out for the first time, you need to give your workplace a classic touch as much as possible. The guidelines below will guide you through setting up your blacksmith workspace.

1. Consider your available space: There is no standard size of a blacksmith workplace; your setup will depend heavily on the amount of space you have available. So before you start drawing plans, consider the space you have available.

2. Choose iron and concrete construction materials: Wood is a bad choice as it can easily be burned by

103

fire. Use concrete and iron as your construction materials.

3. Doors: Determine the size you want to make your door. Make sure your door is wide enough to allow passage of wide and large equipment.

4. Ceiling: Your ceiling's height should be about 8.5-10 feet to allow for easy ventilation and reduce the heat's impact.

5. Ventilation and lighting: Build your workplace in a way that grants you access to natural lighting and ventilation. Make sure your windows are very sizeable and large.

6. A shelf or hanger for your tools: You could create a shelf or wall hanger for your tools or buy an anvil that has one.

Basic Blacksmithing Forging Techniques

Traditional blacksmiths had one conventional technique of forging almost all metals. Most of them have been mentioned in chapter one, so I will give a more expository explanation here.

Drawing: Drawing is usually done to make the workpiece thinner, longer and or wider. To achieve this, place the metal in the forge and position it so that you'll be able to heat only the part you want to draw. Make sure it turns bright red before you pull it out using a tong and place it on an anvil. Then hold it firmly and begin to hammer the metal. The metal will spread as the hammer lands on it. Repeat the hammering process until the metal is all spread out and as thing as you want it to be. Don't hammer it once the metal gets cool; it could break and send all your efforts to waste. However, if the metal gets cold before you are done, reheat it in the forge and resuming hammering until you get your desired result.

Upsetting: This is the thickening of a metal by refolding it. The metal is heated then folded. Note that it won't fold perfectly but bend over; you are to hit it continuously and effectively with a hammer until it fully laps over and becomes thick. You repeat this process until the metal is as thick as you want it to be.

Welding: As explained in chapter one, this is the joining of metals together. This is a very long process under blacksmithing and most smiths weld or substitute it with simpler welding methods using gas and electricity.

However, if you are using a local forge, there are four basic steps to welding:

1. Clean the surfaces of the metals you intend to weld together

2. Heat the metals separately

3. Join the surfaces of the metal together

4. Forge the pieces together

Punching: This is a technique used to make a hole in the metal. To achieve this, heat the part you want to punch till it is glowing red, then take it to the anvil. Place it over the pritchel hole and place the punch over it and be careful not to hammer it on the anvil. That can break the punch and scar the anvil. You wouldn't want that. There are different shapes of punches; go for the one that you prefer to use.

Bending: Bends are easier and more effective when the metal is hot. A metal can be bent manually or by using a spreading tong. When bending the metal over the anvil, place it over the anvil. Let the part you want to bend be lying out, then strike it gently to avoid crushing it. The metal will respond by bending to the pressure of your metal.

Summary

Now we have looked extensively at all you need, from the tools to the safety rules, down to forging techniques you need to put in place.

These are basically what you will need to make your first blacksmith project. So, endeavor to hold onto them as we take a lift into the next chapter.

Chapter 5

Crafting Blacksmithing Projects

As a beginner blacksmith, you have to be careful to pick out projects that are not complicated. Do not be in a hurry to try out fancy projects like the sword or survival knife; such projects are quite complicated and can frustrate you. By choosing less complicated projects, you'll be allowing yourself to start from scratch and build your skills. This way, you will be able to effectively put to practice the knowledge you have gathered from the past chapters. Some of these projects might try your patience before any other thing. In such cases, patience is a more important skill than any other thing you must have gathered; your patience is even worth more than your skill.

Below is a long and well-detailed list of fifteen simple blacksmithing ideas you can try as a beginner. We are using coal forge as a constant in all project ideas.

S-Hooks

This is the number one simplest project in blacksmithing. A great option for a start as it teaches you how to bend, straighten and twist a metal. The s-hook is a tool used in the home for hanging various items.

To start, get a 3/8bar steel. This steel is very thin and straight.

Your forge, hammer, vise, chisel, and tong should be ready and handy. To begin, heat the tail end of the steel in the forge. Once you have it glowing hot, bring it out with your tong and begin to strike using your hammer. After every strike, rotate the steel and strike again to extend the point until it is well tapered.

Once the end is tapered, reheat the rod and make the edges rough round. This can be achieved by continuous rotating and striking of the taper occasionally.

Now, your metal should have a rough round shape and a well-tapered end; reheat it and bend it into the U-shape using your vise and tong. Place your metal in the vise and use your tong to hold and bend until it is well bent and the tapered end is curled over itself. Use your chisel to cut out the excesses. When you are done, slowly quench the hot metal in a cold water or oil to change the properties of the iron and harden it. Don't be in a hurry to get your metal cold and throw it into a bucket of water. It could completely shatter your work

Coat Hook

This is a simple tool found in every household used for hanging coats. You can sell it to friends and family or simply make it for yourself.

You would need a 9-inch piece of round stock (it could be more), a punch as well as other basic tools.

Heat the two ends of your rod in the forge and strike them flat. Once the metal is cold, make a punch mark on it and reheat. Bring out the heated metal using your tong and punch through the mark you have created. Finally, just like you did with the S Hooks, bend the ends into a simple U-curve and use your chisel to remove every misalignment. When you are done, slowly quench the hot metal in a cold water or oil.

Punch

Remember when I mentioned that you could make your own tools as a smith? Well, this is one project to prove that right. A punch could be made from any sized metal or rod item.

To start, cut down your rod to the length you want your punch to be. You can do this with the aid of the chisel, before heating. After heating, hit one end of the punch until you achieve a taper. Then start rotating and striking the taper repeatedly until it becomes perfectly round. When you are done, slowly quench the hot metal in a cold water or oil.

Fire Poker

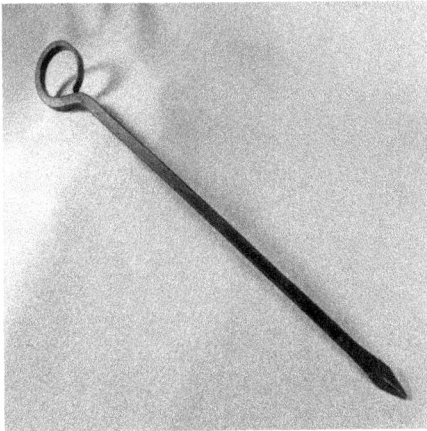

This is another beginner blacksmith project that can be sold for money or used as a gift. It is a tool used to tend fireplaces and is mostly functional in many homes. You would need your forge, tong, anvil and vise to begin. Get a long metal bar and make a thin taper at one end, as we have explained above. Make the taper very long, till the middle of the metal if possible. Then curl the end into a loop. Repeat the heating process and carefully shape the taper until it becomes a round coil. Place the rod in a vise and twist the coiled end around the bar rod. When you are done, slowly quench the hot metal in a cold water or oil.

Bottle Opener

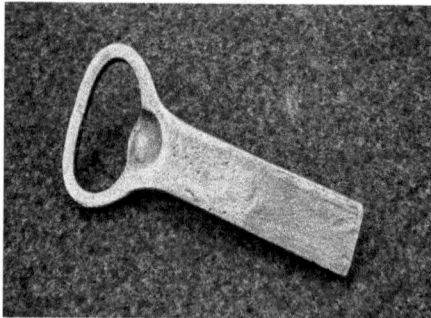

Get your basic instruments together, plus a metal bar of any length. This project is basically simple as it allows you to heat and flatten one end of the bar as we have taught. The next step is to punch that end with a rectangular punch and reheat the metal. After heating, keep hammering the metal until you get a great shape. Use the chisel to cut out the excesses. When you are done, ensure to slowly quench the hot metal in a cold water or oil.

Decorative Heart

This artistic project can be used to make gift items for loved ones or as a display piece. Get a steel bar of ½ inch and your basic tools. Mark the shape of a heart on your anvil. Heat and taper the ends of your metal bar.

Bend the metal at the middle, where the heart will begin from. Then curl the two ends until it meets itself; this will give you the heart shape. When you are done, slowly quench the hot metal in a cold water or oil.

Decorative Cobra

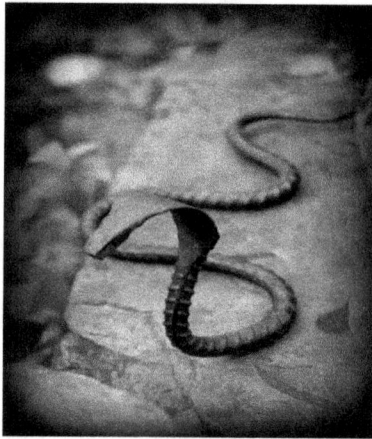

This is an ornamental piece for indoor and outdoor decoration of the door or lawn. To begin, you will need a round stock of considerable size plus other basic tools. You might need a die to help in shaping the head of the cobra. However, you could still make it without the die.

To make the head, taper one end of your bar and bend it into the shape of the snake head. Heat your metal again and flatten a small part of the metal below it to achieve the shape of a cobra's hood. You can curve the other end of the metal to give the appearance of a coiled snake. When you are done, slowly quench the hot metal in a cold water or oil.

Leaf Key Chain

This is quite a challenging project but still involves other basic skills we have talked about in the previous projects, and yes! This is also a decorative item. Get a metal bar and make a tamper of one end into the shape of a leaf. Shape the other end that is not tapered into a straight form, and cut out the excesses. Curl this end over the other tapered end to form a loop as we have in the picture above. When you are done, slowly quench the hot metal in a cold water or oil.

Cold Chisel

This is another blacksmithing hand tool you can make yourself. You will need a forge, a hammer, an anvil, a cold spring, and a chisel for cutting. Get your iron spring and flatten it. Forge the ends of the iron and hammer it to keep it straight and prevent it from spreading out. After you have tapered the end, make the flat end sharp by continuously hitting and rotating it until you have a flattened end. When you are done, slowly quench the hot metal in a cold water or oil.

Froe

A froe is a tool used to cleave woods. It has a very long and straight blade. To achieve this, get a steel iron and heat it till it is red hot. Hammer the steel on the anvil until it is flat and rectangular. Keep hitting it until you get the desired strength and length you want. You can reheat and repeat the process. When you are done, slowly quench the hot metal in a cold water or oil. Finally, stand the metal horizontally and hammer a punch into one end to create space for a handle. Hammer a wide and round enough space as you deem fit.

Meat Skewers

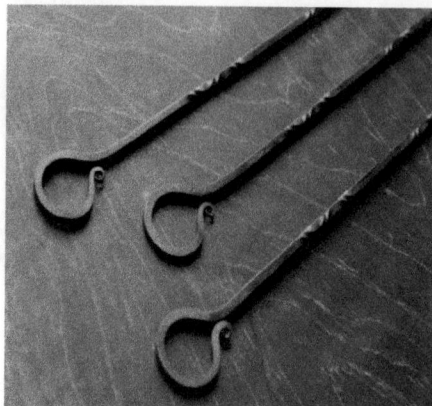

This is made with wrought iron. Heat and taper one end of the iron. Keep rotating and striking the metal until the tapered end has a sharp mouth. Using your vise and tong, curve the end into a half coil. When you are done, slowly quench the hot metal in a cold water or oil.

Nails

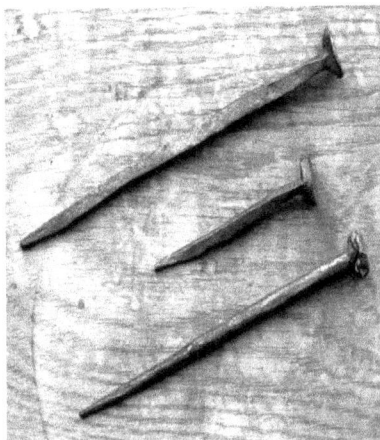

This is quite a technical project as it demands that you make one end very sharp and piercing. Nails are made from iron rods. The first step is to heat the iron till it is glowing red, then rotate and strike until you have a sharp end.

Cut the rod to a very short height and reheat it. Then stand it on the anvil, holding it with a tong, hammer the top slightly to make it flat. When you are done, slowly quench the hot metal in a cold water or oil.

Circular Plate

Get a stainless steel and heat it on the forge. Taper it and use a chisel to cut it round until you have a perfect circular shape. You could use a die to make it hollow. When you are done, slowly quench the hot metal in a cold water or oil.

Dinner Bell

Get an iron rod and mark it in two places where you are likely to make the triangular bend. Heat in the fire till it's hot; it doesn't necessarily have to be glowing hot. Use your tong to bend it at the marked ends. Use another smaller iron to form a circle and weld it to the top of the triangle. To make the circle, heat and taper the iron slightly. Then fold it gently till both ends join together. Reheat and hammer. When you are done, slowly quench the hot metal in a cold water or oil. The dinner bell is simply an iron that is hit to announce dinner, so this should serve.

Trowel

You could make a simple trowel by getting steel metal and tapering one end till it is very flat. You could taper it into a trapezium, triangular or rectangular shape. Once you have achieved this, the other end should be lifted by heating. Heat the metal a little in the forge, then use a hammer to hit it until there is a little lift like we have in the picture above. When you are done, slowly quench the hot metal in a cold water or oil. You can use it this way or hire a woodworker to help you with inserting a wooden handle.

Summary

Yes, even as a beginner blacksmith, you can make very simple household products and sell them in the local market or to people in your neighborhood. That is the strongest motivation that will push you to go further and learn other skills in blacksmithing. This is why I included simple household projects you can sell for money.

I believe you are motivated to move onto the next stage of your blacksmithing venture, which is more learning.

There are still many grounds to cover; if you have covered these grounds, I believe there are no grounds you cannot cover.

Don't forget our golden rule, if you don't get it right, try again!

The end... almost!

Hey! We've made it to the final chapter of this book, and I hope you've enjoyed it so far.

If you have not done so yet, I would be incredibly thankful if you could take just a minute to leave a quick review on Amazon

Reviews are not easy to come by, and as an independent author with a little marketing budget, I rely on you, my readers, to leave a short review on Amazon.

Even if it is just a sentence or two!

Customer Reviews

★★★★★ 2
5.0 out of 5 stars ▾

5 star		100%
4 star		0%
3 star		0%
2 star		0%
1 star		0%

Share your thoughts with other customers

Write a customer review

See all verified purchase reviews ›

So if you really enjoyed this book, please...

>> Click here to leave a brief review on Amazon.

I truly appreciate your effort to leave your review, as it truly makes a huge difference.

Chapter 6

Blacksmithing Frequently Asked Questions

Below is a long list of frequently asked questions in blacksmithing.

Is blacksmithing dangerous?

Yes. It can be very dangerous and requires a lot of safety measures, as we have listed in this book. The fire and tools could be harmful if not well handled. A hot steel is also as dangerous and very hurtful.

Can I use barbeque briquettes as a stove for blacksmithing?

A barbeque briquette can only take as much as 2000 degrees heat, which is enough but not always ideal. Some projects might require more heat, i.e., beyond 2000 Fahrenheit of heat, which your briquette can't provide. Also, they have a lot of chemicals that could melt by high temperature and cause bristles on the steel you're working on. Not an ideal choice.

How much does it cost to start?

That depends on your location, majorly. There are just a few basic things you'll need to get as a starter; they have been listed for you in chapter three. They should be quite affordable. However, if you are low on cash and feel ready to start, you could craft some of the tools yourself. This is one beautiful thing about blacksmithing; you can make your instruments yourself! Of course, you could create a local forge by using coals, coke and little woods. The foot section of a railroad truck is a great option for an anvil. You need something hard and reliable, that should serve.

Where can I get tools from?

A craigslist is your best bet. You could also check scrap yards and antique shops. Make sure to shop locally to get them at a fair rate; online orders are quite expensive. You could also engage in tutorials on how to make these tools yourself. This will do you lots of good as some blacksmithing tools are very expensive and sometimes hard to get. Making them yourself will save you money and help you build your skills. We have explained how to make a local forge, your cold chisel and various types of punches. Ensure to give these projects a trial and engage in further learning to make them appear more presentable. Learn the tricks to make other tools too.

How do I know the size of anvil I need?

Your budget.

The kind of project you intend to work on.

How much blacksmithing you plan to do.

It's totally dependent on you. You can buy the biggest size if your budget can accommodate that. Also, the kind of project you intend to work on determines the quality of tools you need. You'll surely need a three hundred pounds' anvil for a huge projects like machine parts. Also, if you intend to go into full-time blacksmithing, you'll need something large to last you a long time. However, if you are low on funds and don't intend to do heavy projects, a 150 pounds is a good start. There are still other sizes in between you could go for to suit you best.

Can women blacksmith?

Yes, a woman can blacksmith. It is very interesting to find a lady doing blacksmithing. There are a few ladies already in the industry. Blacksmithing is not only about strength but also involves a high level of creativity and patience. Once she is creative and patient, she is good to go. All she needs is to master the process and use her safety tools. Anyone can blacksmith except children.

How long does it take to be a professional?

No timeframe can be used to answer this question; it takes as long as the efforts you input, the right efforts to be precise. Being a professional is no fast flight, but the knowledge stays with you forever when you finally become a master. You need a long time of continuous practice of several projects. You can become a professional in less than a year. It is all up to you.

However, if you want to study a course, it could take you two to four years.

What to do to become a professional?

Like every other art, blacksmithing can be mastered. You have to be hardworking and patient enough to try one project several times till you get it right. Consistent practice of blacksmithing is the pattern of the masters; you don't wait till you have major projects before you hit the steel. Even when there is no project, try to get steels and try out professional projects; you could gift it to your friends, family and loved ones.

Never allow your steel to gather dusts; take NO BREAK. Long breaks in between practices will reduce the ratio to which you would have mastered the art. Finally, you could also train with professionals or study it as a course. If you are seeking to practice this trade for the most part of your life, be ready to go to extreme lengths and make sacrifices to make yourself better. Nothing great is cheap, by the way.

Can you learn on your own?

Oh yes! There are self-help books you can get from the library to learn on your own. You don't necessarily need a standby teacher. There are also YouTube videos to watch practical tutorials. You could visit blacksmithing websites for further learning; there are quite a number of them. However, if you have the chance of having a live training, it'll go a long way in helping you.

How can I focus heat on one small part of my metal?

An acetylene torch is what you need. Smiths from the 18th century would have given anything to have this hand tool available to them. It can produce a heat temperature of about 5260 degrees. It uses acetylene gas plus oxygen to generate temperature. When you buy it, the gas tank will be empty, which will require that you refill it. Also, it is very versatile and can be handled by anyone! A great tool for a beginner it is. It saves you the dirt, stress and hassle from the conventional forge. It can be used to cut, weld and braze any kind of metal.

It is also a brilliant idea for designs as you can reposition the metal so that you only get to put heat where needed. It makes decorative twists much easier.

Do I need to be strong to be a blacksmith?

You averagely need endurance techniques and creativity, not a mass of strength. If you have strength but lack all of these things, you will not do so well. Most persons think that being a blacksmith requires mostly energy. They think it is about the hammer's strength and pressure coming down in contact with the metal, but you need to know, like we have mentioned several times in blacksmithing, you need to learn the right amount of pressure. If you are angry and need to punch some bag or release some energy, the forge is not a place for that; turn to your gym or any other activity that could help you release good energy. You can't heat a project with hatred and anger and expect it to turn out fine; it doesn't happen. Blacksmithing is not a craft of just muscles.

Do anvils get damaged?

Yes, but it will be quite a long time before that happens. After a long period of use, it could get pierced, open up and damaged. There is a 12% chance of the anvil getting damaged. You can use rectangular 4*4 metal boxes, ensure it has 200 pounds below it, so it doesn't move about and a 70-80° rebound to save you some energy when you hit your iron. Anvils made with cast iron are a poor choice. They could peel off and cause your metal to get rough. Also, anvils made with cast iron do not last long. It takes just one heavy hundredth strike to break it in. These anvils are usually sold cheaper than the other types, which makes them more appealing to purchase. Do not exchange low value for money because it is cheap. It is best to save for a better and stronger anvil or to use a strong makeshift.

What is the best holding tool that I can get?

Tongs, vises, and clamps are unique holding hand tools; they are not exactly the same. A tong is a special versatile tool with two grips at the end and a string in the middle. It's used for picking and holding things out of the forge. A clamp is a class for strengthening and holding things together. A vise is box with two parallel jaws for holding metal while the smith works. It could be permanently placed on a bench; however, it can be used for the same purpose; holding a metal workpiece. There are different sizes of these tools that serve different purposes. A good smith has various shapes of these tongs and vises for special projects. The tools you use for a flat metal can't be used for a round one. And what a 1/2 tong can carry, a 1/4 tong cannot carry it. Samples of different tongs are listed in chapter four.

What tools are used to cut a metal during blacksmithing?

Cuts are made on metals by hammering the punch or chisel on the metal. The place that needs to be cut should be carefully marked out first. Then the smith positions the chisel on the mark and begins hitting the chisel against the metal until it gives way and the process continues until the cutting is done. If you intend to use punches, ensure they are square mouthed. Round punches are not a great choice for cutting, except you want to mark out a round shape.

There is also an instrument called hardy. Hardy has a sharp end like a chisel, but its head is like that of the hammer. It is very heavy and has a long handle. Hence, if you seek to use this tool, you wouldn't need a hammer. The pressure it can generate is directly equal to that of a hammer. It is used for cutting heavy and thick metals like machine parts and iron rod. Some smiths use the hammer to hit on the other end as the hardy hits the metal. Two persons can do this style. One person holds the hardy and the other holds the hammer. Use this method when you are dealing with heavy metals only. Hardy could be used for cold and hot cutting.

However, cutting is very much easier in blacksmithing once the metal is hot. It could be very stressful when the metal gets cold. Metals like copper and stainless are very much easy to cut in comparison with other metals. If your project requires cutting, make sure to hit it until it is hot, not necessarily red hot. If it is too hot, it could break during hammering and your efforts would go to waste. If it is too cold, the metal could also break. Just ensure that the project is hammered at an even temperature. During the cutting process, the metal should be held by a sturdy vise to get a perfect cut. The wrong vise could throw your metal to the floor or give you shaky hands. So the basic tools for cutting a metal is a punch or chisel plus a hammer and a sturdy vise.

Are there special tools for designing?

The basic and general tools used for designing are tongs, chisels, anvil, forge, punches, and hammers. These tools are mostly used together to make specific shapes. For example, to make a round shape, the metal is heated up and placed on a vise where a tong is used to curl and bend it until it forms a circular shape. The same method applies to most designs. Nonetheless, there are special designs that can not be made with just these tools. Designs like the artistic forging of different flowers on a gate or plate. This kind of blacksmithing requires induced pressure. Some machines have been designed to meet these needs.

Examples of such a machine is the induction forge and it involves the use of electricity to heat the metal.

What are the most technical projects in blacksmithing?

There are several projects that could cause you to sweat like never before and keep you standing and forging for hours. Most of these projects are sword and knives. The technicalities involved in making these projects is not something a beginner can handle. There are basic skills that need to have been mastered. Any mistake made during knife or sword making could mar the whole project. The smith must go through training before attempting a knife or sword. The attention that the blades require cannot be given with basic knowledge.

Is there any law guiding blacksmithing?

Not really. It depends on your location. In most places, there are no blacksmithing laws. The smith is only guided by the organization or association he belongs to. Asides from this, there is no particular law attached to blacksmithing. However, farriers (explicitly applied to a blacksmith whose specialty is in shoeing horses) have an official organization in certain locations, and it is unlawful to practice as a farrier without being registered with them. Also, all blacksmiths must comply with the health and safety, and fire legislation. This is where most smiths encounter problems because they fail to keep to the laws of fire safety established and when the fire safety officials come for a routine inspection, they get fined them and lock up their workshop. We have already discussed how to set up your workplace, follow the guidelines we have listed and ensure your safety equipments are handy.

Also, general legislation is required, like a sales tax license, contractor's license, and a UL or CA number. The sales tax license enables you to sell your products in public markets without being harassed. The contractor's license licenses you to install architectural work. The UL or CA number is necessary if you are into the making of wire lighting items. None of these things is existent in one place, register for the one that is valid in your locality to avoid harassment. However, a business license is operational in every country; try and get that as a basic license.

Is it necessary to buy an anvil stand?

No, not at all. There are makeshifts tools in blacksmithing. You can use a strong and well-balanced wood or iron. What is important is that it is sturdy enough to carry the anvil and the metal you will be forging.

At what age should a person stop blacksmithing?

Old age is what will definitely come to anyone whether we prepare for it or not. And at such times, health begins to fail. A person of sixty years should retire from blacksmithing for health's sake. I know some persons are still strong at that age, but it is not appropriate nor safe for an older person to be beside constant heat and making hits.

What is the best heat treatment to give to a metal?

There are four different types of heat treatment. All of them can be used on a single job, depending on the type of project.

Annealing is for softening carbon steels by slowly heating it in the forge then burying it in ashes to prevent it from cooling too fast. The longer it takes to cool, the softer the metal. This is a softening process.

Normalizing is a heat treatment given to a metal at the end of the forging process. The metal is heated to its highest temperature then allowed to cool naturally.

Hardening is the thickening of medium carbon steels. This process works well with this type of steel. It is achieved by heating the metal to medium temperature then quenching with water or oil. Oil is a great option, but in the absence of oil, water can be used in mixture with earth or salt. Once the metal gets cold, it becomes harder. There is no need for hammering.

Tempering is the heating and flattening of a metal to increase the toughness and reduce the hardness.

So, I cannot tell you the best heat treatment to give to a metal; it is totally dependent on the type of project you are working on and the effect you want to achieve.

How do I make a bellow for my coal forge?

There are different DIY bellows for a coal forge. The most reliable and easy type is the goat skin bellow. The toughness of the skin and durability make it a perfect choice. Goat skins have been used over the years for several projects in Africa.

To make the goat skin bellow, get a big female goat (the skin of a female goat is always softer than the male skin after it has been treated). Slaughter the goat and hang it on a tree your hand can reach. On that tree, peel off the skin using a knife. Peel it carefully, so it comes out in whole without breaking at any point. The skin should be turned inside out and hung under the sun. Tie the two ends from which the legs were peeled. Then pour sand or stone into it.

After a while, scrape the skin. This scraping should be repeated for a couple of days. Then the skin should be oiled with groundnut oil. The next process is beating the skin with sticks until the leather is soft and very flexible. Again, be careful not to cut it. Now, add a pipe to one end of the legs and tie it together. Also, fix or sew word sticks over the top opening. Still leave the other hole tied. So when using the bellows, the top part is opened for air to come into the bag and closed afterward so that the air is pressured to pass through the connected pipe into the fire. Your goat skin bellow is ready! Other types of bellows are rotary bellows, Chinese box bellows and water bellows. If you cannot make the goat skin bellows, then you should try water bellows. This is a recent and modern type of bellow and a perfect replacement for the goat skin. It is easier and less stressful. However, it could only be made where there is an available old oil drum.

What is cold forging?

This is an unpopular form of forging and it is not popular amongst old blacksmiths. Traditional blacksmithing involved the use of direct fire and no other methods, but in recent times, several blacksmiths have started to adapt to this method for decorative purposes.

In cold forging, the stock is placed inside a die. Dies have been explained in chapter two. This stock is squeezed against a second die. The stock is allowed in there at room temperature for a long while until the deformation process is completed. This process requires high-duty steel with very strong resistance. The type of steel required for cold forging is low carbon steel with carbon content below 0.5%.

How do I make sales and advertise myself as a blacksmith?

This is one issue that scares a lot of persons from the idea of blacksmithing. Worries like 'Who will buy my products? Who will I keep blacksmithing for? How do I reach a larger market?' are not uncommon today. With machines producing neat and basic home tools, the blacksmith seems to have lost his place. However, there are little tricks to getting recognition and making sales as a blacksmith. I have a little trick I call, Show Tell and Sell. You can call it STS for short. This trick begins with you practicing various project ideas. Take one idea or two at a time. Practice making your project idea, give it your very best and read extensively before you commence. If it turns out fine, good for you! If it doesn't, go back to the heat and try again until you get it right.

When you indeed get it right, do not be in a hurry to sell. Give that steel as a gift to a close family friend or an influential person in your locality. The latter is a much better choice for the sake of the trick. Do these for about three to four times to three different persons and tell them to help you tell others about your business. One out of them would definitely comply and tell others, and even pay for some specific projects. This is what the show, tell and sell strategy is about. You show your expertise, then encourage others to tell others and then you roll into sales.

Another means of STS is sending your work to museums or places mostly visited by people; you could also place it by a busy roadside or a popular park, just anywhere that allows you to show it off. Let the workpiece do the telling while you sell. Method one is more expensive and effective than this method. However, I bet both will give you more sales than all your signboards will ever bring. In chapter one, we talked about how social media can help you modernize your blacksmithing business. You could create a blog or activate your account so that you can reach a wide range of people before you try an ad. Ads will be more effective if you are visible and ensure to upload beautiful pictures of your work all the time.

How do I correct mistakes during blacksmithing?

And this is the last question we will be reviewing. I like this question because it allows me to resound what I have said earlier and get comfortable making mistakes. There are some mistakes you can never really correct. For example, breaking a rod of iron. When you break an iron you intended to bend, the next thing that comes to your mind is to weld it and try again. If you succeed in welding it, you cannot bend that metal again. So, that is a mistake that cannot be corrected. Mistakes that you can correct is coiling the wrong part of the metal. You could uncoil it and straighten it again. If you mistakenly soften the metal, you can also harden it with heat treatments. There other mistakes you could also correct, but it will take you more time and cause you more stress. One favor you can do yourself is getting comfortable making mistakes and facing the consequences of your mistake.

Conclusion

You'll notice that throughout most of the chapters, one word was consistent, basic and simple.

We ensured to make use of the simplest terms and techniques you can relate to.

It is possible to have seen a blacksmithing project somewhere you want to replicate and don't know how. Well, this is just a map and a guide as well as an introduction for you into the journey of blacksmithing.

It doesn't in any way limit you from exploring other projects that I said earlier or trying other tricks out. All we did was to equip you with basic and foundational knowledge on what blacksmithing is about. So, you can ride on the wings of creativity and explore other tricks and projects in blacksmithing.

I am so rooting for you. I believe that by following the letters in this book, you will be able to do well. These are just tidbits that can help you achieve your goal of being a freelancer in the world of blacksmithing.

You are holding a guide and a manual; you could always refer to it for clarification. That is why we have explained everything to you in simple terms.

You might also need to read books by professional smiths to help your advancement if you intend to kick start a professional blacksmithing career. There are a lot of them in the local stores. As I have said several times, this book is just for beginners and should have equipped you with the fundamental knowledge you need to help you start.

Go to the nearest garage shop close to you, and begin to purchase your tools and materials. You can create your workshop yourself, but if you have issues doing that, get a carpenter and start practicing right away. If making the workshop will be challenging, your backyard will be a good place to start.

So, why don't you start now?

Happy blacksmithing!